Look there! Monkeys!

MARVELS ANIMALS 07

MONKEYS

KATE RIGGS

CREATIVE EDUCATION | CREATIVE PAPERBACKS

4

table of contents

Published by Creative Education and Creative Paperbacks
P.O. Box 227, Mankato, Minnesota 56002
Creative Education and Creative Paperbacks
are imprints of The Creative Company
www.thecreativecompany.us

Design by Graham Morgan
Art direction by Blue Design (www.bluedes.com)

Photographs by Flickr (Biodiversity Heritage Library), Getty (Kevin Schafer, photomaru, Roy Toft/National Geographic Creative, silviacoluccelli, viafilms), iStock (Kittisuper), Pexels (Aleksey Kuprikov), Shutterstock (Cynthia Kidwell, Daleen Loest, Eric Isselee), Unsplash (Cassie Gallegos, kalai venthan gopal, Syed Ahmad), Wikimedia Commons (Muybridge, Eadweard)

Library of Congress Cataloging-in-Publication Data

Names: Riggs, Kate, author.
Title: Monkeys / by Kate Riggs.
Description: Mankato, Minnesota : Creative Education and Creative Paperbacks, [2025] | Series: Marvels | Includes bibliographical references and index. | Audience: Ages 4-7 | Audience: Grades K-1 | Summary: "An engaging introduction to monkeys, this beginning reader features eye-catching photographs, humorous captions, and easy-to-read facts about this tropical animal"-- Provided by publisher.
Identifiers: LCCN 2023046961 (print) | LCCN 2023046962 (ebook) | ISBN 9781640268449 (library binding) | ISBN 9781682773949 (paperback) | ISBN 9798889890126 (ebook)
Classification: LCC QL737.P925 R54 2025 (print) | LCC QL737.P925 (ebook) | DDC 599.8--dc23/eng/20231109
LC record available at https://lccn.loc.gov/2023046961
LC ebook record available at https://lccn.loc.gov/2023046962

Printed in China

Monkeys are animals that are good at climbing. Most monkeys live in trees.

Monkeys have long arms and legs. They have long tails, too.

Some monkeys have brown or black fur. Other monkeys are white or gray.

MMMM, BANANA.

Monkeys eat plants or fruit. Some monkeys eat bugs or birds.

YUM!
YUM!

WHAT IS UP THERE?

A young monkey is called a baby. A baby monkey lives with other monkeys in a **troop**.

Monkeys groom each other. Then they look for food. Monkeys like to climb and play, too.

BAD FUR DAY.
SO MESSY!

SEE YOU LATER,
ALLIGATOR.

Farewell, monkeys!

[Picture a Monkey]

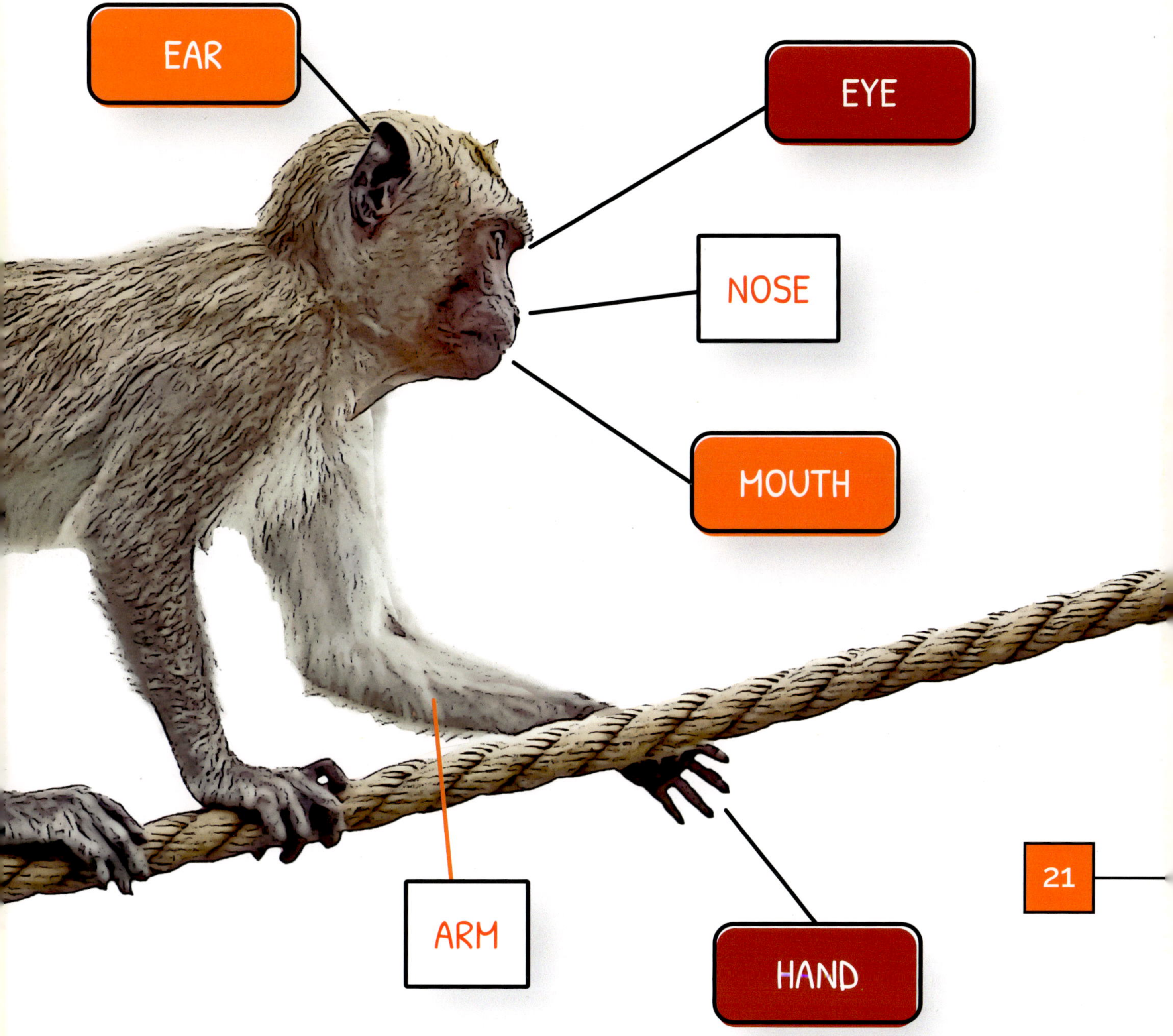
EAR
EYE
NOSE
MOUTH
ARM
HAND

WORDS TO KNOW

groom: to clean face or hair

troop: a group of monkeys that live together

READ MORE

Lang, Suzanne. *Grumpy Monkey Don't Be Scared*. New York: Random House Studio, 2023.

Thielges, Alissa. *Monkeys*. Mankato, Minn.: Amicus Learning, 2024.

WEBSITES

Learn to Draw a Monkey
https://artprojectsforkids.org/draw-a-monkey-2/

Easy steps to draw a monkey and a monkey coloring page.

Monkey Art Projects and Activities
https://www.dltk-kids.com/animals/jungle-monkeys.html

Try monkey coloring pages or make a monkey mask!

INDEX